Early praise for *Getting Started with Software Defined Radio*

This book serves as a great introduction to beginners because it collates all the requisite hardware and software tools. It also helps you walk through the SDSharp and play with the tool.

➤ **Sai Yamanoor**
   IoT Applications Engineer, Buffalo, NY

The manuscript was very well written, and I've been having fun playing with the AIRSPY software tuning in stations and listening to digital police scanners in my county in Georgia.

➤ **Michael J. Lewis**
   Technology Enablement Consultant, Slalom Consulting LLC

I could recommend the book to someone who is interested in a practical approach to radio signals but has little to no experience on them. The book is an approachable introduction with step-by-step instructions for installing and running software.

➤ **Oona Räisänen**
   C++ Programmer, Founder of windytan.com

Well written and very comprehensive, it is a good introduction to a hard topic.

➤ **Gianluigi Spagnuolo**
   Reverse Engineer, Exein

Explore Software Defined Radio

Use SDR to Receive Satellite Images and Space Signals

Wolfram Donat

The Pragmatic Bookshelf

Raleigh, North Carolina

Many of the designations used by manufacturers and sellers to distinguish their products are claimed as trademarks. Where those designations appear in this book, and The Pragmatic Programmers, LLC was aware of a trademark claim, the designations have been printed in initial capital letters or in all capitals. The Pragmatic Starter Kit, The Pragmatic Programmer, Pragmatic Programming, Pragmatic Bookshelf, PragProg and the linking *g* device are trademarks of The Pragmatic Programmers, LLC.

Every precaution was taken in the preparation of this book. However, the publisher assumes no responsibility for errors or omissions, or for damages that may result from the use of information (including program listings) contained herein.

Our Pragmatic books, screencasts, and audio books can help you and your team create better software and have more fun. Visit us at *https://pragprog.com*.

The team that produced this book includes:

Publisher: Andy Hunt
VP of Operations: Janet Furlow
Executive Editor: Dave Rankin
Development Editor: Patrick DiJusto
Copy Editor: L. Sakhi MacMillan
Layout: Gilson Graphics

For sales, volume licensing, and support, please contact *support@pragprog.com*.

For international rights, please contact *rights@pragprog.com*.

ISBN-13: 978-1-68050-759-1
Book version: P1.0—February 2021

Contents

Acknowledgments

The offspring suggested I make a few acknowledgments: "I'd like to thank my arms for always being at my side, and my legs for always supporting me..."

I'd like to think I'm not *quite* that dorky, though I may be close. Instead, I'd like to thank:

Chloe, for ensuring that all projects built are equipped with evasive maneuvering capabilities.

Oliver, for making sure that the office door hinges work well.

Loki, for ensuring that I'm able to hear in case of an emergency and for making sure I'm not forgotten in the office.

Smudge, for emotional support (both giving and receiving).

Sai Yamanoor, Mike Lewis, Oona Räisänen, Gianluigi Spagnuolo, and Youssef Touil for tech editing the book, spotting errors, and ensuring that the projects worked for everybody. What errors may remain in this book are mine and mine alone.

Everyone at The Pragmatic Bookshelf for taking a chance with this book.

Patrick Di Justo, for doing an awesome job, as always, of making my stuff sound the way I *wanted* it to sound.

Rebecca and Reed, as always, for putting up with my disappearances and weird projects, sometimes with little or no explanation.

Introduction

For several years in my travels online, I kept hearing and reading about SDR, or software-defined radio. It seemed interesting, but (at least at first) I didn't have the time or motivation or learn more about it.

I gradually picked up a news item here or a Reddit post there, but I still didn't know the details of SDR. All I knew for sure was that the topic was complicated and it allowed you to use your computer to pick up radio signals. Instead, I concentrated on more easily learned things like partial differential equations and making the perfect soufflé.

Then, not too long ago, I got an email from my editor. "Hey Wolf," he wrote. "How would you like to do a book on SDR? Ever heard of it?" It was time to jump into the pool again, and this time I didn't come out until I figured out just what the heck was going on in this weird mix of hardware, software, Internet, and radio waves and wrote it all down.

This book is the result. It's all about how to use your computer or laptop to pick up radio signals from the earth and space using a fairly recent technology called SDR.

On the surface, receiving radio signals doesn't seem like a big deal. After all, everyone's familiar with satellite TV, right? Even before Dish Network and DirecTV made it commonplace, anyone with a few grand and a good view of the sky could erect a 2-meter satellite dish in their backyard, hook it up to their TV with an esoteric jumble of electronic bits and pieces, and enjoy watching Japanese television in their living room. The radio waves are there; it's just a matter of receiving and decoding them.

In 1901 Marconi showed everyone how to communicate across the Atlantic using a high-powered device that produced electromagnetic waves—the first radio transmitter—and a huge antenna. Ham radio operators (named because professional radio operators thought these amateurs were so bad at tapping out Morse code, they must have hams where their fists should be) have been listening to signals for more than a hundred years. You can listen to NOAA's

Weather Radio on any FM radio that can access the lower end of the VHF (Very High Frequency) band, as well as any marine VHF transceivers and a number of weather radios sold commercially. You can tune into local police frequencies and hear everything from standard police chatter to forest fire information to breaking emergencies. You can listen to FM and AM broadcasts in your local area. I have a shortwave radio in my garage that can pick up stations around the world if the weather is right. And satellite TV receivers are now commonplace, to the extent that they compete with cable television in many areas.

But in our world, inexpensive commercial radio receivers are usually limited to one or two segments of the radio spectrum. I can purchase a police scanner, and learn how to use it and what frequencies are being used by law enforcement agencies in my local area, but the police scanner may not be able to pick up the NOAA weather broadcasts. And the weather radio probably can't tune in the ham radio frequencies. And a ham radio probably can't get my local Top 40 station. Wouldn't it be great if there was an electronic device you already owned that could be trained—dare we use the word *programmed*—to tune into different frequencies all across the radio spectrum?

Enter SDR.

SDR, or software-defined radio, is the technique of using a small receiver—most often a repurposed USB TV device—to tune in and listen to radio broadcasts at various frequencies. Advances in microcircuitry and software make it possible for many of the functions of multi-band radios to be handled programmatically. The result? You can use your laptop to listen to the police scanner, or local radio stations, or even (and this is where it gets really cool) download and see satellite images from various satellites, including some of the NOAA weather satellites.

What exactly is needed to do all of this cool stuff? At the bare minimum, you'll need a computer with a sound card or other ADC (analog-to-digital converter), an SDR unit (usually a USB dongle), and an antenna. Thanks to advances in technology, assuming you already have a desktop or laptop computer with fairly standard capabilities, you can pick up the necessary SDR equipment that will enable you to listen to quite a wide range of signals for less than $50. You can also use a Raspberry Pi to do all of this, meaning that an entire SDR setup can be built for under $100.

Most computers and laptops have an onboard sound card, and higher-end ones may have a stand-alone ADC board, depending on what the computer is being used for. In my experience, the standard integrated sound card on even a low-end laptop is perfectly capable of processing the signals correctly.

The USB dongles used are usually those designed to receive and decode high-definition digital television broadcasts, though the slowly rising popularity of SDR and its growing number of enthusiasts has led to several of these devices being specifically made for—and marketed to—the software-radio crowd.

The antenna is the final piece in the SDR puzzle that can cause some headaches: which antenna do you use? What shape does it have to be? How big? And where do you point it? I'll go through each of these questions and a few of the possibilities available to you when it comes to picking out or building an antenna in the chapter on antenna theory, but you may be comforted by the fact that you can use an old-school set of rabbit ear antennas without too much modification being necessary.

Software-defined radio itself is pretty easy with today's technology; in fact, I think that the most difficult thing about it is figuring out exactly what you can do and what you can't based on the equipment you may happen to have. Because so much of each SDR installation is custom made—you select the software you will use, which antenna, which USB dongle, and so on and so forth—it can be difficult to match your configuration to a known-good working configuration. If your setup matches another exactly except for the version of software being used, it's entirely possible that your setup will fail while the other one will receive all sorts of signals without a problem. A lot of information is available on the Internet, but there seems to be a scarcity of getting-started guides that walk you through the process from start to finish. In addition, the information that is out there is widely scattered. If you're interested in getting started with this interesting hobby, you're stuck reading five-year-old blog posts, poring through subreddits, and trying to interpret poorly written documentation for software packages that stopped being actively supported back in 2013.

I hope this book helps you, the reader, to find your way through the maze of information out there, figure out what exactly you want to do, and show you how to do it with a minimum of fuss and cursing—I did that part for you, at least. Ready? Let's get started!

Materials Needed

In this book, aside from technology and software, I use a few different bits and pieces. I thought a list of the things I use in one place might be useful to you before you get halfway down the rabbit hole and realize you're missing two important things you need *right now* to finish a project. To that end, here's what I use in the book:

An SDR USB dongle, https://www.amazon.com/gp/product/B009U7WZCA/

An extra antenna for said dongle, https://www.amazon.com/gp/product/B013S8B234/

Extra long antenna cable, https://www.amazon.com/gp/product/B00685RFC2/

Coax cable adapters, https://www.amazon.com/gp/product/B072JCR57H/

A pair of rabbit ears (an antenna, that is—no rabbits were harmed in the making of this book), https://www.amazon.com/gp/product/B000EIMKYC/

A flower pot

Some PVC pipe

CHAPTER 1

Installing the Required Bits and Pieces

So you've decided to explore the wonderful world of software-defined radio, the world of virtual transistors, analog-to-digital and digital-to-analog converters, and homemade antennas. For that, I say congratulations and welcome! I also say good luck and Godspeed, for here be dragons. I have buckled my armor and girded my loins for the express purpose of guiding you through the turbulent waters of turning your computer into a radio receiver. *And if that isn't a mishmash of metaphors, I don't know what is. Anyway...*

With the proper SDR tools, including software, tuning device, and antenna, you can use your computer to tune into a large swath of the radio spectrum, from 64 MHz—the lower part of the VHF bands—all the way up to the 1700 MHz UHF bands and beyond. There are tools (Airspy HF+, for instance) that will tune into the spectrum all the way down to 1 KHz, and still others, such as HackRF and LimeSDR, that will go higher than 1700 MHz. These are a bit on the expensive side and perhaps beyond the scope of this introductory book, but know that they do exist should you want to explore more of the fringes of the radio spectrum. Instead of turning a tuning dial to change radio stations, as you do with your car radio, with SDR you merely tell the software to tune the device to a specific frequency. This can be done with astonishing levels of precision; some software will allow you to increment or decrement your scan in units of 0.00001 MHz. That's well beyond what even the most precise, steady-handed person can achieve with a manual tuner. As you can see in the image from http://www.transportation.gov, SDR doesn't cover a huge portion of the radio spectrum, but having one device able to tune into that many frequencies is very impressive.

Given the right antenna and software, it's often possible to use the same USB dongle to listen to FM radio stations, CB radio (see sidebar), police and fire scanners, the International Space Station, and even to tune into and visualize the signals from NOAA weather satellites.

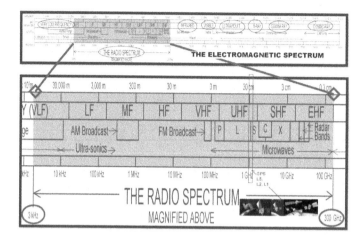

Once the signal is acquired, it's routed from the antenna, through the tuning hardware (most often a USB dongle of some sort) into your computer's sound card. The sound card acts as a digital-to-analog converter, taking the analog signals and converting them to digital signals which your computer can pipe to the speakers. Some signals, such as those from weather satellites, have an image embedded into the signal being transmitted. With the correct software, you can extract and view those images.

Getting started with software-defined radio can be a challenging experience. The SDR world is still a fledgling area with little documentation, despite numerous subreddits, websites (http://www.rtl-sdr.com, for instance), and online/Facebook groups. As I write this, Hackaday is even hosting an online chat with SDR guru Harold Giddings, who goes by the call sign KR0SIV. SDR may be growing in popularity, but it's still kind of hackerish.

As such, what few SDR software programs exist are often buggy, platform-specific, and—most of all—poorly documented. Dozens of hours of research can lead to nothing but confusion and frustration. When I was getting started, the frustration was often palpable; there's nothing like following a long set of instructions, step by step, for over an hour, only to find when you finish that it doesn't work and it's anyone's guess as to why.

One thing I learned through my travails and experiments is that it's important to be pragmatic when it comes to the tools—including the operating system—you decide to use to play with SDR. I'm primarily a Linux guy, for example; if you've read any of my other works, you'll know that I am most

Citizens Band Radio Service

In the United States, Citizen's Band, or CB, is a two-way, short-distance voice communications service operating near 27 MHz in the shortwave band. It can be used for both personal and business messages. It probably reached its height of usage in the 1970s and early 1980s, but the availability of pagers and, later, cell phones have returned the service to its original users, long-haul truck drivers and hobbyists. Younger readers may not be familiar with it, but just watching an old movie like *Smokey and the Bandit* may introduce you to nostalgic phrases like "10-4, good buddy" and "chicken lights" and "There was a plain brown wrapper at the 60-yard stick, a bear in the air, and a wreck at the 405. The coops were workin' hard on your side going west."

The original CB specs called for AM transmission, but over time channels 36 through 40 became used for SSB communication. To listen to CB conversations, tune to one of the MHz frequencies below on AM. From 27.365 MHz and up, use either LSB or USB.

- 26.965 | 26.975 | 26.985 | 27.005
- 27.015 | 27.025 | 27.035 | 27.055
- 27.165 | 27.175 | 27.185 | 27.205
- 27.215 | 27.225 | 27.235 | 27.245
- 27.255 | 27.265 | 27.275 | 27.285
- 27.295 | 27.305 | 27.315 | 27.325
- 27.335 | 27.345 | 27.355 | 27.365
- 27.375 | 27.385 | 27.395 | 27.405

When you're comfortable with the tools discussed in the book, try listening in and see if there's any CB traffic in your local area!

comfortable in Ubuntu and Raspbian. I do, however, use both Mac and Windows as well, as I'm not a purist and will happily switch to whatever tool is best for the particular job I'm doing. And I almost always try to keep my books OS-independent, giving instructions for all three major OSes.

However, when it comes to SDR, Windows is still the OS of choice, so we'll often be using that when it comes to the projects in the book. Linux definitely has some software out there, and I had great success with some of it, but unfortunately, it seems that many of the Linux tools are very hardware dependent; the same piece of software may work fine on your desktop system, but switch to one with a different USB chip and it all may fail. In addition, some tools are just easier to download and use in Windows than in Linux.

As for you Mac junkies, OSX (or MacOS) now has many of the same tools available that Linux has, including rtl-sdr, airpsy, and hackrf. GQRX, the main Linux tool I use in this book, is also available from either its website or via homebrew or macports. However, while I was writing this book, I stayed firmly in the Linux and Windows arenas, so I don't know how much success you may have with these tools or how easy they may (or may not) be to install. Just know that they exist, and if you try them out and they work well, please let me know!

Hardware

Let's talk about the hardware that's necessary for any software-defined radio experimentation. The first thing you're going to need, obviously, is a radio—or its equivalent in the SDR world: a USB dongle. Most dongles in the SDR space have been originally designed as TV tuners, to allow the user to receive HD TV signals out of the air. As their popularity has grown (for both tuning into television signals and for software-defined radio enthusiasts) they've come down significantly in price, and many of them can be used to detect all sorts of signals, given the right antenna. Repurposing these dongles fits one definition of hacking: making a device do something it wasn't built to do.

The most common SDR dongles you're likely to see use the RTL28xx interface and the Realtek R82xx tuner chipsets, housed in a variety of different packages. The dongle I'm currently using (see the image that follows) is from NooElec and is available from your favorite online retailer for around $20 (https://www.amazon.com/gp/product/B009U7WZCA/).

Others are available, of course, from various retailers, so don't feel any pressure to purchase one over another. That being said, however, remember my earlier warnings about getting different setups to work? If you're completely new to the SDR world, it may behoove you to duplicate my efforts here exactly, starting with the hardware I'm using. I would hate for you to duplicate my steps exactly but have the project not work because of some vague mismatch between your hardware and your software, or because your USB device isn't readable by your SDR software.

The next piece of hardware you'll need is an antenna, to grab all of those beautiful signals out of the air and funnel them into your SDR dongle. Chapter 4 is all about antenna design and theory and which antennas will do the best job for particular projects and signals, but when you're just getting started and getting familiar with the processes involved, you just need any old antenna.

The NooElec dongle in the previous link comes with an antenna; my experience is that the included antenna is worth about as much as a snowblower in the Mojave. Two hours spent trying to listen to the local radio station and failing to get anything convinced me to try another antenna I had picked up when I was still unsure as to what I needed. Don't be afraid to switch antennas, as switching antennas can make all the difference, along with placement (which we'll get into later as well). When you're first getting started and are just trying to pick up some signals—any signals—you may have good luck with this one: https://www.amazon.com/1090Mhz-Antenna-Connector-2-5dbi-Adapter/dp/B013S8B234/ (see the following image). I certainly did. Out of all of the pieces in the SDR puzzle, the antenna may make the most difference. You may get your setup to work perfectly, but if your antenna is wrong (such as being the wrong design or having the wrong placement), you may have no luck picking up signals. Feel free to experiment.

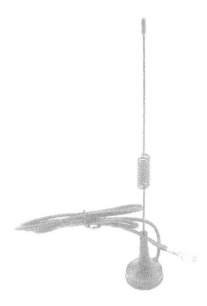

We'll be switching up our antenna for our later projects, but this is a good one to start with. Whichever one you choose, make sure that the connector matches the connector on your dongle, which is most likely an SMA (the first of the two images that follow) or an MCX (the second image). Happily, many add-on antennas come with an array of adapters to fit most any radio device.

SMA male

MCX male

MCX female

You will most likely want to get a longer cable. The stand-alone antenna I bought, for instance, comes with a 1-meter cable, which is plenty for some simple experimentation—picking up your local radio station, for instance. However, success with SDR depends not only on the antenna but the antenna *placement*. Getting the antenna far away from your computer and other noisy devices is crucial, especially as the strength of the signal you're trying to receive decreases. As I said, we'll get into antenna design a bit later on, but a longer cable is almost guaranteed to be a necessity. Again, make sure the extension cable you choose fits not only the antenna but the connector on your USB dongle. Also make sure your genders are correct on each end of the cable; you may want to purchase a selection of gender-changing adapters to go with your SDR toolbox.

That's the bare minimum of hardware you'll need to start experimenting. Read on for an introduction to the software we'll be using.

Software

Ah, the software. Here's where things can get a bit sticky, so bear with me while I attempt to lead you through the jungle. In a nutshell, you'll be finding and installing a new device driver and a tuning program. Sounds simple, right?

Using your USB dongle for SDR experiments requires, at a minimum, a device driver that is more adaptable than the standard manufacturer's or Windows or Linux drivers. Once you get the new driver(s) installed and working, you'll also need software that enables you to tune the dongle to your choice of frequency. This is often called RTL-SDR software. Windows users most often use a program called SDRSharp, while Linux users tend to use a package called GQRX.

Windows

For simplicity's sake, all the Windows work you see here will be done on the latest build of Windows 10. It's *likely*, though not guaranteed, that the packages you see will also work with Windows 7 and 8. A newer computer is also a good idea, but anything as powerful as a dual-processor or better should be fine.

SDRSharp, the most common Windows program, requires Microsoft's .NET version 4.6 or newer to be installed. If you're using Windows 10, this may already be installed, but don't count on it—I tested this process using a brand new install of Windows 10 Home edition and .NET 4.6 was missing. You may also need the Visual C++ runtime. (Don't worry—you won't be doing any C++ or .NET programming; those libraries are just necessary to compile and run the software.)

To install .NET, go to https://www.microsoft.com/en-us/download/details.aspx?id=55167 and choose your language from the pull-down menu. Click the big red Download button and follow the instructions to install it. The Visual C++ package is very similar; go to https://www.microsoft.com/en-us/download/details.aspx?id=8328 and again choose your language. Click the Download button and follow the instructions to install the package.

When the installation is finished, you should now have all of the operating system tools you'll need. Now, point your browser to https://airspy.com/download. Click the Download button next to Windows SDR Software Package (as shown in the image on page 8).

When the download is finished, you'll have a file named sdrsharp.zip in your Downloads folder. Move that file to a directory of your choosing and extract it by right-clicking and selecting Extract Here. You'll end up with a new directory called sdrsharp-x86. (Don't worry if you're running a 64-bit system; the software will install and run just fine.)

Navigate inside that directory and run install-rtlsdr.bat. Make sure you're connected to the Internet before you run the .bat file. A command prompt should open, which will then attempt to download two new files to your directory: rtlsdr.dll and zadig.exe. *Both* of these files are necessary to proceed. The file rtlsdr.dll is the modified driver for your SDR dongle, and zadig.exe is a handy tool for telling Windows to use the new driver instead of the old one (because the old one won't work for SDR experiments).

Although zadig.exe downloaded for me without any problems, rtlsdr.dll did not. If this happens to you, you'll need to download that file manually. Point your browser to http://osmocom.org/attachments/download/2242/RelWithDebInfo.zip, which will put a new RelWithDebInfo.zip file in your Downloads folder. Unzip the file the same way you did the sdrsharp.zip folder, and you should have a new directory named rtl-sdr-release.

Navigate inside that folder to either the x32 or x64 directory, depending on what version of Windows you're running (most likely 64-bit). Copy this new rtlsdr.dll file into the sdrsharp-x86 directory.

If by chance the zadig.exe file didn't download correctly, enter https://github.com/pbatard/libwdi/releases/download/b730/zadig-2.5.exe into your browser window and copy the resulting .exe file into your sdrsharp-x86 directory.

That finishes off the software you need to download. Now you need to run everything. Plug in your dongle and wait for Windows to try to find or install drivers for it. Don't worry whether it succeeds or not, since you'll be replacing those drivers in a moment.

When it's finished, open your sdrsharp-x86 directory, right-click the zadig.exe file, and select Run as Administrator. This will open the following window:

Zadig is a nifty little tool that lets you choose what drivers you want to use for a particular device—in our case, the SDR dongle. In the menu bar, select Options and make sure that there's a check mark next to List All Devices. Also, uncheck Ignore Hubs or Composite Parents to make sure that you can see everything connected to your computer.

Now, in the main drop-down menu, you'll need to select your dongle. It should appear in one of two ways: as Bulk-In, Interface (Interface 0), or as something like RTL2832UHIDIR or RTL2832U. Choose whichever one shows up in your menu. You'll know you have the right device when the USB ID showing is 0BDA 2838 00. Oddly enough, every tuner dongle will have this USB ID, no matter the manufacturer. Do *not* select anything else, because this can severely screw up your USB drivers.

Underneath the drop-down menu, you'll see a box labeled Driver, which is prepopulated with whatever driver Windows happened to select for your dongle. In the box next to it (which the big green arrow is pointing to), make sure WinUSB (vX.X.XXXX.XXXXX) is selected. This is the driver you're going to use to replace the Windows default. Click the big blue Replace Driver button.

You're almost guaranteed to get a warning about unverified publishers and unsigned drivers; just ignore it and install the software anyway.

You should now have the drivers necessary for your SDR dongle to work. It's possible that if you unplug your dongle or move it to another USB port, you *may* need to run zadig.exe again, so don't delete it from your sdrsharp folder.

On to the next chapter for your first radio reception.

Linux

If you prefer working with the penguin, you're in luck, since there is SDR software out there for you, though it's not as easy to install and use as AirSpy is for Windows. I'm using Ubuntu, because it's what I'm most familiar with and its distribution is most widely supported in the SDR community. If you're using Fedora, RedHat, Debian, or something else, your results may vary. I'm using Ubuntu 16.04, but feel free to use the most recent release which, as of this writing, is 19.10.

We'll be using both the rtl-sdr package from the Ubuntu repos and a program called GQRX. To install the rtl drivers, simply open a terminal and enter

sudo apt-get update

sudo apt-get install rtl-sdr

After installing the package, you'll need to see if your distribution is using the DVB-T drivers, which some have loaded by default. To check, enter

sudo rmmod dvb_usb_rtl28xxu

in your terminal. If you get the response rmmod: ERROR: Module dvb_usb_rtl28xxu is not currently loaded, you're golden. If, on the other hand, you get a different response stating that the command was successful (or no response at all, which also means it was successful), you'll need to permanently disable the drivers, since you'll be replacing them with something else and the rmmod command is only temporary until you plug in the dongle again or reboot.

To disable them, create a file called rtlsdr.conf inside the /etc/modprobe.d directory (you'll have to do it as sudo). In that file, enter

blacklist dvb_usb_rtl28xxu

and save it. Reboot your computer and you should be good to go.

To test that everything is working so far, open your terminal and enter

rtl_test

at the prompt. You should be greeted by something similar to the terminal screen on page 11.

If you're unable to get the rtl_sdr packages working, you may need to install the drivers from source. To do that, first install the necessary dependencies:

sudo apt-get install libusb-1.0.0-dev git cmake

Then clone the rtl-sdr repo:

git clone git://git.osmocom.org/rtl-sdr.git

Once it's cloned, make a build directory, build it, and install it:

cd rtl-sdr

mkdir build

cd build

cmake ../ -DINSTALL_UDEV_RULES=ON

make

sudo make install

sudo cp ../rtl-sdr.rules /etc/udev/rules.d

sudo ldconfig

When it's finished, follow the preceding instructions about the Linux DVB-T drivers and then test with the rtl_test command.

Now you can install GQRX. As I said, I'm using Ubuntu, and the easiest way to install it is using the aptitude package manager (apt-get. However, you'll have to add the correct repositories to your sources.list file first. To do that, enter the following in your terminal:

sudo add-apt-repository -y ppa:bladerf/bladerf

sudo add-apt-repository -y ppa:myriadrf/drivers

sudo add-apt-repository -y ppa:myriadrf/gnuradio

```
sudo add-apt-repository -y ppa:gqrx/gqrx-sdr
```

```
sudo apt-get update
```

Follow that with

```
sudo apt-get install gqrx-sdr
```

You should now be able to open it by typing

```
gqrx
```

in your terminal. Assuming everything installed without issue, you should now optimize it by installing the libvolk1-bin package and running the volk_profile tool by entering

```
sudo apt-get install libvolk1-bin
```

```
volk_profile
```

It'll take a few minutes to run. When it's finished, on to the next chapter for your first radio reception.

CHAPTER 2

Your First SDR Reception

Now that you've got all of the software and drivers installed, it's time to connect your dongle to your antenna, open up the software, and see what signals you can receive. Again, I'll split the reception part into two sections—one for Windows and one for Linux.

Attaching the Antenna

Once your USB dongle is plugged into your computer and the drivers are installed, you can connect the antenna. Using either the cable that came with your antenna or the extension cable that you got, attach it to the SMA or MCX connector on the dongle, making sure it's secure. Then run the cable to the antenna, placing the antenna as far from your computer as you can (remember, even for your first experiments, you're trying to eliminate as much noise as possible, and your laptop or computer is quite noisy, radio-wise). If you can place the antenna near a window, so much the better. I had problems keeping the antenna upright, but a little double-stick tape on the base of the antenna solved that problem quickly.

Once the antenna is attached and secure, go back to your computer and start the software.

Windows

Windows users are going to use the SDRSharp application to tune the dongle to different frequencies. It's in the sdrsharp-exe directory you created and modified earlier in the book, so if you don't have that directory open already, open it now.

Volume versus Gain

In the following sections we're going to talk a lot about *gain* and *volume*. So what exactly is the difference between the two? If you increase the gain on a device that has a knob for it, you probably notice that the sound gets louder, which is exactly what happens when you increase the volume.

In a nutshell, volume adjusts the *output* level in decibels of your device. Gain, on the other hand, adjusts the *input* level decibels of whatever channel you have selected. Gain controls how loud something is before it goes through any preprocessing, preamplifying, or any other pre- something or other. It will affect how loud the signal comes through the speakers, but in the case of SDR or other similar processes, you're amplifying the signal as much as you can to isolate it from the background noise. RF gain is adjusted in the hardware before the signal is even digitized and is independent of the operating system.

There's little difference in how the Windows and Linux platforms handle gain and volume, with one notable exception: the *noise floor*. The noise floor is the level of the background noise around the signal—the background static that forms the "floor" of the spectrum that you're seeing on the graph. In my experience, a value of about 1.5 dB seems to work pretty well in the Linux configuration. On Windows, a gain of around 30 dB is what works best.

Double-click SDRSharp, and you should be greeted with the window shown here:

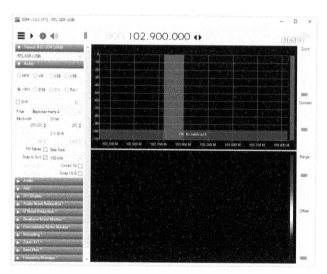

There's a lot of stuff in this window, so let's take a moment to familiarize ourselves with it. At the top center, as you've probably guessed, is the frequency in megahertz to which the program is currently tuned. Underneath the frequency label are two graphic representations of the frequency spectrum in the general neighborhood of the current frequency. You'll see those in action shortly. The sliders along the right side of the window allow you to adjust what you see in those windows, including the contrast of the displays and how much of the spectrum appears in the window.

The left side of the window contains the fine-tuning settings you'll be playing with as you experiment. You can choose the source of the signals, what form they're in, and play with things like gain, filters, bandwidth, and any number of other settings that we won't be getting into at the moment.

For our first experiments, we're just going to try to pick up a nearby FM radio station. First, you need to tell SDRSharp where the signals are coming from. At the top left of the window you'll see a drop-down that allows you to select the source. Select RTL-SDR (USB) from the choices. The first time you choose it, you may get a warning from your computer about how Windows is protecting you from evil files and devices. If this happens, click the More Info button and then choose Run Anyway.

Next you'll want to adjust the gain of the signal, because if it's too low you won't be able to receive anything but static. Click the gear icon at the top left, above the Source selection drop-down, and you'll see the small menu like that in the following image.

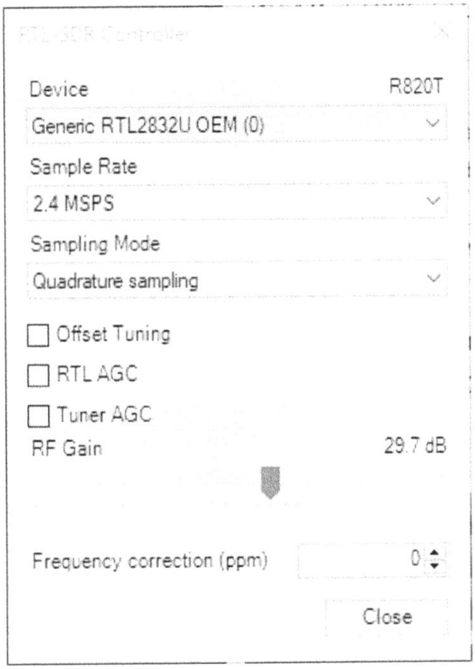

You can leave the other selections at their default settings, but move the RF Gain slider to somewhere around 30 dB, and then click Close. For those of you following along with both Windows and Linux tools, SDRSharp's RF Gain slider is a bit different from the Gain slider in GQRX. In SDRSharp, this slider changes the RF hardware setting, while the slider in GQRX changes the volume within the software.

With your gain adjusted, you're ready to tune in. Click the Play button (the triangle) next to the gear icon at the top left of the window, and you should be greeted with the sound of static coming through your computer speakers and what looks like a seismograph scrolling toward the bottom of the screen (see the following image).

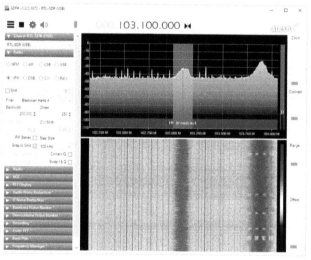

As we stated earlier, the numbers across the top of the window are the frequency to which you're currently tuned; in the case of the previous image, it's tuned to 103.1 FM. This happens to be a station in my local area, and you can tell there is a decently strong signal because of the "hump" in the background frequency graph. Also in the window pictured, you can see that there's another strong signal between 103.75 and 104.0 FM. As you may have guessed, there's another local station at 103.9 FM in my area. This is a good illustration of how the top graph shows the relative signal strengths of the local "neighborhood" in the RF spectrum; strong signals show up as peaks above the surrounding static. The stronger the signal, the higher—and most likely the broader—the peak will be.

Of course, it's unlikely that you're tuned to a local station when you first bring up SDRSharp, so you'll need to look around in your local spectrum—the equivalent of spinning the tuning dial on your old FM radio. There are a few ways of doing that. For larger tuning adjustments, you can click the individual digits in the frequency display; clicking the upper half of the digit will increase it by one, and clicking the lower half will decrease it.

Once you're in the neighborhood of the frequency you're looking for, you can fine tune your adjustments by either individually adjusting the digits (changing 103.100.000 to 103.101.000, for example), or by clicking the vertical red line in the center of the frequency graph and dragging it to the left or right.

As you scan through the spectrum, I'd suggest two things to keep in mind when you're looking for signals. First, don't overlook the sliders to the right of the frequency window, especially the Zoom slider. You can use it to navigate precisely within the section of frequencies that you're scanning.

Second, experiment with the strange-looking icon directly to the right of the frequency display in digits (it looks like two triangles pointing at each other). When the icon looks like that, it's in Center tuning mode, which means that the red tuner line will always be in the center of the window. Clicking it again will put it in Free tuning mode, which will allow you to move the red line within the window, rather than keeping it in the center. This can be useful if you're trying to zero in on and investigate a particularly small or elusive frequency spike.

Before you know it, you should be tuning in to your local radio stations and even picking up some that are further away. Experiment with your antenna as well; try moving it around and pointing it in different directions. If you're having difficulties, skip to the Troubleshooting section after the Linux section of this chapter.

Linux

Hopefully you had no problems getting the rtl-sdr drivers and GQRX installed. Once your antenna is connected, you can start receiving signals by opening a terminal and entering

gqrx

You may get a crash-warning window, which states that the current settings are not optimal and may cause the program to crash. Click Open Anyway. You should be greeted by the window shown on page 18.

Like Windows' SDRSharp, there's a lot of info here, so take a moment to look around before you blindly start clicking and dragging (unless you're like me and that's just how you do things). The main part of the window, on the left,

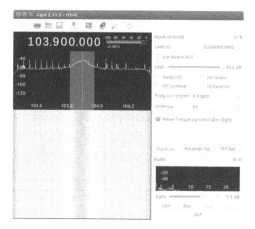

displays the current frequency in megahertz to which you're tuned, with a gain meter just to the right. Below that are the graphic representations of the surrounding neighborhood of the RF spectrum.

The top of the window contains the standard File menu options, like Preferences, Open, Save, and so forth. The right side of the window is where you perform all of your advanced settings and tunings, like antenna selection, choosing hardware, adjusting gain, and choosing filters.

The first thing you'll need to do is choose your receiver as the input device. Click the icon on the top row—directly above the spectrum image—that looks like a PCI-E board or a circuit board.

In the resulting window, select Realtek RTL2838UHIDIR SN:00000001 (or the device most similar to that, depending on your hardware) from the Device selection menu. Leave the other options as their defaults and click OK.

Next, you'll probably want to adjust your gain (see previous sidebar), because otherwise it's very easy to lose your signal in the *noise floor*. In my experience, a value of about 1.5 dB seems to work pretty well. In GQRX, the LNA slider is equivalent to the RF Gain slider in SDRSharp, so play with both of these sliders until you're comfortable with the results.

Finally, choose the frequency to which to tune your device. Similar to Windows' SDRSharp, you can tune the frequency window by clicking the individual digits in the shown frequency; clicking the upper part of the digit increases it, clicking the lower part decreases it. You can also click and drag the vertical red tuner line in the middle of the spectrum graph to adjust it by hand. You can also zoom into the frequency using the Frequency Zoom slider in the FFT Settings tab.

When everything is set, click the triangle-shaped Play button in the top row. Before you know it, you should be tuning into your local radio stations and even picking up some that are further away. Experiment with your antenna as well; try moving it around and pointing it in different directions. If you're having difficulties, check out Troubleshooting in the next section.

Troubleshooting

If you're having issues, hopefully this section can help you narrow down the possible causes and fix the problem.

You're not hearing anything, or you're hearing nothing but static, even at frequencies where there should be a signal (like your favorite local radio station).

This is the most common issue people seem to have. The answer to this is probably the antenna. First, make sure it's firmly connected to the dongle. If that doesn't fix things, either move the antenna further away from your computer (which may require a longer cable) or switch to a different or better antenna. If you only have the one antenna, try simply picking it up and pointing it in different directions. If you start picking up a signal, even a very faint one, you'll know the problem (and solution) is the antenna.

You're getting a No Device Selected or No Compatible Devices Found message when you start SDRSharp.

This is a common problem in Windows. These messages can sometimes be caused by low-quality USB cables or hubs. You can try moving the dongle to another USB port, or removing it from a hub and plugging it directly into the computer. You may also try plugging it into a USB 2.0 port rather than a USB 3.0 port. I had the best luck in this case with simply starting over and reinstalling all of the drivers.

It's still not working.

Starting over seems to be a viable option. If you decide to go this route, uninstall or delete everything you've done so far. If you're using Linux, this will include the command apt-get remove and all of the packages you had to install via apt-get previously. As I said, these programs can be touchy, and if the wind is blowing in the wrong direction or your socks are the wrong color, the installation may fail completely. Keep trying and I guarantee it will eventually work.

When it does, turn to the next chapter for some antenna theory, and then to the first project!

Try This

Some things to try when you've had some successes and are feeling comfortable/cocky:

1. Try using a new antenna setup to receive commercial radio stations outside your general geographic area. Do you have better luck during the day or at night? How does the weather affect reception?

2. If you're using a laptop, try making your setup portable and travel for a bit to a different area, preferably one with a different geography (for example, if you live in a valley, try going up a mountain). Do you have better luck? Does altitude affect reception?

CHAPTER 3

Antenna Theory and Design

It's an amazing kind of sorcery that you can point a piece of metal at the sky, attach it to some various electronic bits and pieces, and suddenly listen to a radio broadcast from the other side of the planet.

That piece of metal pointed at the sky is the antenna. As you may have guessed, when it comes to receiving radio signals from space (or from anywhere else, for that matter) the kind of antenna—the shape and size, and the direction in which you point the antenna—is very important. Before we get into the various types of antennas that work best for SDR, I'd like to briefly discuss how antennas work.

We could talk about the physics of antennas for an entire book, and there are entire textbooks and courses of study about the subject. But that would be way outside the scope of this book, and unnecessary. This chapter is just meant to give you an introduction to how antennas operate for both sending and receiving. We'll briefly discuss how antennas transmit a signal, why you need different kinds of antennas to *receive* those signals, and what kinds might work best for the particular projects in this book.

How Antennas Transmit

To begin with, all radio transmissions are electromagnetic (EM) waves. To transmit a radio wave, a moving magnetic field induces a current and a voltage in a length of wire. You can test this yourself by connecting a sensitive voltmeter to a loop of wire and wiggling a strong magnet back and forth within the confines of the loop. You'll see a small voltage and current appear on the voltmeter. It's a small current—on the order of several microamps—but that's because you're not wiggling the magnet fast enough. If you increase the frequency of your magnet-wiggling, the voltage and current will increase in the wire, which in turn will increase the power of the transmission (remember your electrical equations—$P = IxV$, or power equals current times voltage).

Now, let's pretend that your transmitter is just wiggling a magnet *really* fast around some wire. This creates an *oscillating electric dipole* (a dipole is a system where the positive and negative charges are separated by a distance). The oscillation creates an electromagnetic wave that emanates from the antenna (or the wire, in this case). The following image shows a dipole antenna emitting an electromagnetic wave in the same plane as the two antenna wires.

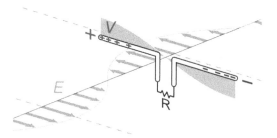

As the magnet wiggles, electrons in the wire are moved from one end of the dipole to the other, creating a wave. That wave travels through three-dimensional space in pulses that can be displayed in two-dimensional form as a sine wave.

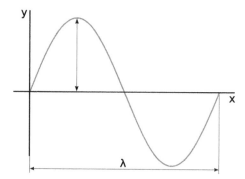

Since all electromagnetic waves travel at the speed of light in a vacuum, or $3x10^8$ m/s, the frequency of the pulses distinguishes one type of wave from another. (Yes, the speed of light is sliiiiiightly slower in air than in a vacuum, but I don't think the difference between 299,792,458 m/s and 299,702,547 m/s is worth fixing, do you?) Frequency is measured in cycles per second; one cycle per second is called one Hertz (named after the German physicist Heinrich Hertz). The frequency of those waves varies according to the kind of transmission; the color blue—specifically, blue light—has a frequency of around 630 THz (terahertz), while the radio station just down the road from me transmits at a frequency of 103.1 MHz (megahertz).

As you might infer from the examples I just gave, there's really no difference between a color of light and a Top 40 radio station transmission, electromagnetically speaking. They're both just different frequencies of EM waves. Higher, faster frequencies may manifest as visible light; lower frequencies can manifest as television signals, radio signals, or even sound.

The other side of the frequency coin is wavelength. As you can see in the first image above, the wavelength of a wave is the distance between peaks (or troughs) of that wave. Since all EM waves travel at the same speed, the frequency of their pulses determines their wavelength. Higher frequency pulses mean a shorter wavelength; slower frequencies mean longer wavelengths. The equation that combines the two is pretty simple: electromagnetic wavelength is defined as the speed of light divided by the frequency of the wave. This means that for an FM radio broadcast of around 100 MHz, the wavelength is about 3 meters. A broadcast frequency of 10 GHz, on the other hand, has a wavelength of about 3 centimeters.

You may also hear or read things about the *bandwidth* of a signal. The bandwidth of a particular signal is the difference between its highest and lowest frequencies. For instance, if a signal is transmitted between 100 MHz and 120 MHz, its bandwidth is 20 MHz. Signals with a tight (small) bandwidth often have more power but are more difficult to "catch" out of the air, as you have to more finely tune your receiver.

Another difference between different signals is the concept of a waveform. You may wonder, if two signals share the same range of frequencies in the electromagnetic band, what keeps them separate and distinguishable from each other? The answer is waveforms. Waveforms are definitely outside the scope of this book, but in general, think of waveforms as different frequencies of waves added together to make a unique shape. Thus, one signal may have a certain range of frequencies included within it, while another neighboring signal may look totally different. This keeps them unique within a part of the EM spectrum, while allowing a certain type of antenna to capture both. The top image shown on page 24 illustrates this concept; the bottom right of the image is an example of two waves "added" together to form a new waveform.

The final piece of the transmission puzzle (that interests us, anyway) is the polarization of a signal. If you have a single piece of wire as your transmitter, also called a monopole, then the electromagnetic wave emitted from the antenna is matched in orientation, or polarization, to the orientation of the

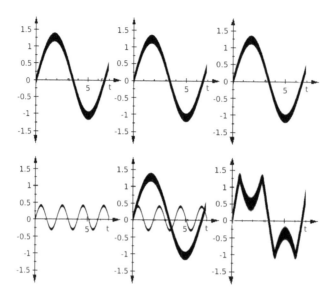

antenna. (A monopole is a single piece of wire, while a dipole is two separate pieces of wire that are parallel and lie almost end-to-end with each other.) What this means is that if your transmission antenna is sticking straight up and down, your receiving antenna (assuming it's also a monopole) needs to also be straight up and down to get the best reception. Linear polarization, such as in this example, isn't very useful; what good does it do to transmit a signal if receivers have to be oriented the exact same way in order to receive it? A good example of this is shown in the image that follows, where the blue (horizontal) wave is polarized 90 degrees from the red (vertical) wave. In this instance, a horizontally oriented antenna would be unable to receive the red signal, while a vertically oriented antenna would be unable to "see" the blue wave.

So what do you do to ensure that receivers can always receive the transmission, no matter what direction their antenna is oriented? In the case of linear polarized signals, one way is to use two antennas oriented at 90-degree angles to each other. This ensures that no matter how the receiver is positioned, one of the antennas will be getting a strong signal. (One could also argue that you could mount a series of single antenna wires around a single base, each slightly out of phase with the previous one, so that all angles are covered.)

While this will work, and will cover your bases when it comes to orienting the antenna, this method isn't normally used, because the additional weight of all of those added antenna wires would make the end product unwieldy and unstable. Instead, a better way to combat this problem is to change the polarity of your transmission signal. The most common types of signal polarity are horizontal, vertical, right-hand circular, and left-hand circular. It's a well-known fact among radio enthusiasts that vertically oriented signals often travel better than horizontally oriented ones when it comes to local communication. However, circular polarity is often used to fix the orientation problem, because most circular-polarized signals are broadcast in all directions at once (depending on the broadcast pattern), making the antenna orientation irrelevant. This type of signal is often used for satellite communications, as it's impossible to make sure that the satellite's broadcast antenna is in the same orientation as the receiving antenna. In addition, the antenna on a spacecraft is often rotating around the axis of the craft, constantly changing direction and orientation. The terms *right-hand* and *left-hand* refer to the direction of the helical turns as viewed from the receiver. The wave in this image is right-hand polarized:

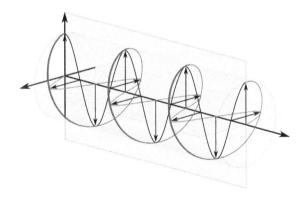

How Antennas Receive

This brings us to talking about how to *receive* signals, and this is where we look at antenna shapes and sizes. As you probably figured out, shapes and sizes of antennas are determined by the frequency and polarization of the signal you're trying to catch.

Let's look at how to receive horizontal and vertical polarization first (the simplest). Just like the broadcasting antenna, these are best received by monopole or dipole antennas.

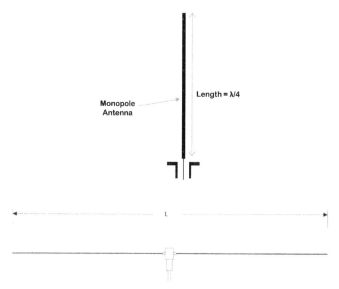

You may notice that the length of the antenna is explicitly depicted in both figures; this is because in both cases, the length of the antenna poles has a specific relationship to the wavelength of the received signal. A monopole antenna will experience the best reception if the antenna is one-quarter the length of the signal wavelength. A dipole, being really nothing more than two monopoles stuck together, will get the best reception if the total length of the antenna (both monopoles measured together) is one-half the signal's wavelength.

Obviously, the length requirement is not a hard-and-fast rule, otherwise the whip antenna on your car (or the one built into your windshield) would be able to pick up only one station—the one whose signal frequency matched the antenna's length. Rather, it means that antenna performance will be optimized at a particular length, and if you are designing an antenna to pick up one particular signal, then taking this rule into account will be to your advantage. An interesting experiment for a budding antenna physicist would be to use a handheld FM radio with a variable-length antenna and see if changing the length of the antenna affects the reception of different local stations. Following the equations given earlier, a monopole antenna should be set to about 75 cm to best pick up an FM station transmitting at 100.1 MHz, and making it longer should optimize reception of lower-frequency broadcasts (99.9, for example). Meanwhile, making it shorter (remember, higher frequencies mean shorter wavelengths) should optimize receiving stations at 101.1 and above.

The final antenna shapes that we'll talk about are those tuned to circularly polarized signals. And what shape is used for that type of reception? Not surprisingly, it hearkens back to the series of antennas mounted circularly around a central mast I mentioned earlier, but in this case it's more of a DNA shape: the *helix* antenna.

The simplest helix antenna is the *monofilar* helix in the image that follows. We don't want to get too deep into antenna physics, again, so just know that things like the radius of the turns, the distance between them, the total length of the antenna, and the pitch angle affect the reception power. At its core, however, the monofilar helix is just a monopole antenna like the whip antenna we discussed earlier, only with an omnidirectional radiation/reception pattern.

The type of circular polarization of the signal directly affects the shape of the antenna. A helix antenna can be either right- or left-hand polarized, and that polarization must match the signal polarization, otherwise the antenna won't receive anything. The next step in the antenna complexity stepladder is the double-helical design, which—just as it sounds—imitates the DNA helix and makes the antenna a bit more effective at capturing signals.

Another, more complex antenna design you may find in your antenna design quest is the *skew-planar wheel antenna*. It's not used very often—mostly in radio control (RC) designs—but it has the unique characteristic of receiving linearly polarized signals no matter its orientation. It's why it's used in RC scenarios; the airplane can still receive signals from the RC transmitter, no matter what direction or orientation it's in.

Although it's a pretty interesting design, it's relatively rare, so I won't go into it any more here.

Antenna Design for SDR Hobbyists

So how does all of this theory affect us SDR hobbyists? The answer, of course, depends on how far down the rabbit hole you wish to go and what sort of results you're expecting. If you've already experimented with the projects in Chapter 2 and Chapter 3, you've most likely been successful with the simple whip antenna that connects to your SDR dongle. You may have also noticed that some of those basic monopoles are built with a small coil in the middle of the antenna. This is simply a way of making the antenna's effective length a bit longer without making it too unwieldy. Luckily, it doesn't affect the antenna's effectiveness (though I have seen some advertisements attempt to market the coil as a reception enhancer—make of that what you will).

Those simple whip antennas will work fine for the digital signal decoding in Chapter 5, as well, but you'll have to switch it up a bit to get a hold of the NOAA satellite signals in Chapter 6. That's because they, like most satellites, broadcast using a circular polarity—a right-hand circular polarity, in their case. Ideally, you should use a right-hand circular polarized antenna to pick up their signals, but luckily some genius in the SDR world discovered some time ago that a pair of rabbit ear antennas could also be used, given that they were spread apart *this* far and oriented just like *that*. I've got more info about how to set up the antenna in that chapter.

In the meantime, I hope this chapter gives you an introduction to just how these magical pieces of metal actually grab sound and pictures out of the sky. Though it's not an in-depth study on antenna physics, it should give you enough of an intro so that if you find a signal you're interested in, you can decide what sort of antenna to build or buy to have the best chance of receiving it. I highly recommend you search for it on YouTube, as I've discovered that videos and gifs make a huge difference when trying to understand some of the concepts, especially the electromagnetic radiation patterns.

If you're comfortable with the concepts, let's move on to decoding some digital speech encoding.

CHAPTER 4

Digital Speech Decoding

Let's revisit, for a moment, the old days before silicon integrated circuits and software packages and newfangled computing machines. Back then, an enterprising hobbyist could use some transistors, a few variable resistors, a diode or two, and a tunable crystal and, with the right antenna and a healthy dose of skill and talent, receive and listen to almost any signal on the air. Back then, all broadcast radio signals were analog. While UHF broadcasting existed, there was almost nothing transmitting in that area of the spectrum, and satellite radio was unheard of.

Nowadays, analog signals are gradually being replaced by digital ones. The day may not be far in the future when all signals on the airwaves are digital. Back in 2009, all U.S. analog television signals were shifted to digital, and in early 2019 the nation of Norway shut off all their national analog FM radio stations in favor of digital audio broadcasting. If and when that happens in your location, the type of radio set shown in the preceding image will no longer be useful. It will still pick up the signals, but they will be unintelligible to the listener.

The difference between those older signals and the newer digital ones is just like the difference between old cassette tapes (remember those?) and CDs. Cassette players work by transforming audio waves back and forth into analog electrical signals, usually by way of a microphone or a speaker. Similarly, analog radio broadcasts work by changing electrical signals into radio waves that are transmitted. The signal directly represents the transmitted sound (or picture) by varying voltages and frequencies.

Digital music storage, such as CDs, on the other hand, works by storing the sound as patterns of ones and zeros, the same way a computer stores data. To listen to a CD, you have to read the pattern of ones and zeros with a laser and then decode them. Likewise, digital radio broadcasts transmit patterns of numbers rather than analog waves. While your SDR dongle can pick up both types of signals, if someone is broadcasting via a digital signal, you won't be able to understand it—at least not without using some special software to convert digital to analog.

A perfect example of this digital encoding is found on the public safety and law enforcement bands. First of all, you should know that contrary to popular belief, it's *not* illegal to own and operate a police scanner—at least in the United States. (Check with your local laws if you're not reading this in the U.S.) You can purchase them on Amazon. However, should you decide to get yourself such a scanner, you may be disappointed, because many of these public departments have made the switch to digital broadcasting.

The reason many public departments have switched to digital is often because it's much cheaper to broadcast a digital signal than an analog one, and *this* is mainly for two reasons. First, digital radio offers better resistance to interference from other signals nearby on the spectrum. This means it can be broadcast with less power behind it, since you don't have to worry about having to overpower neighboring signals. And second, since it avoids the necessary physical imperfections of analog transmitters, more of the power you put into the broadcast gets translated into the signal rather than into heat energy loss in the broadcast equipment. So less power is needed to broadcast, and you get more bang for the buck with the power you utilize.

What all of this means is that if you're interested in listening to your local fire department or sheriff's station, you'll need to be able to decode the digital signal. Luckily, you can do this by adding some free digital decoding software to your SDR suite of programs.

One thing I find necessary to mention: digital speech or signal *encoding* is a completely different animal than signal *encrypting*. Digitally encoded speech signals can be decoded and listened to pretty easily with free software. Encrypted signals, on the other hand, are often official channels and frequencies that are not meant to be decoded or listened to by the general public. While software exists to decrypt these signals, it's often illegal for civilians to use. In other words, stick to your local police department broadcasts and stay away from broadcasts by the NSA or the CIA.

Finally, a word of notice from your author: although I managed to get digital speech decoding up and running on Windows, it was tricky. I had much more luck with my Linux installation, though it may have to do with the difficulty I had in locating suitable DSD-encoded broadcasts. If you're trying this on a Windows box and are having absolutely no luck, you may want to consider taking the leap (if you haven't already) and transitioning to a Linux system, at least for your SDR experimentation.

Hardware

Luckily, no matter which operating system you're using, no additional hardware is required beyond what you were using to listen to standard FM radio signals. That being said, however, you stand a much better chance of receiving some good signals if you have a good antenna. In preparation for listening to weather satellites, it might be a good idea to find or purchase an old TV rabbit ear antenna setup.

These are available on Amazon or your local big box store for around ten dollars. I would suggest that you choose a set that terminates to a coax connection. You can then follow up your antenna purchase with two things: a longer cable for your dongle and an adapter to connect that longer cable to the coax cable on the antenna. Both of these are available on Amazon—look for an MCX male-to-female extension cable (often used for GPS systems) and an MCX-to-coax adapter. Once you've got your antenna and cable setup, you're ready to receive digital signals, as well as the older analog signals.

Software

Once again, I've separated instructions for Windows and Linux. If you're using Linux, feel free to skip ahead.

Windows

In addition to SDRSharp, you'll need both a program called dsd (which stands for digital speech decoding) and a way of sending the output from SDRSharp to the dsd application, via software running inside your computer.

Ordinarily, when you tune your dongle and listen to the radio frequencies with a program like SDRSharp, the output is being piped (obviously) to your computer's speakers. It's really no different than listening to the radio in your car. If you want to decode digital speech signals, however, you need a way of sending the output of SDRSharp to a digital speech decoding program instead, and then sending *that* output to your speakers.

This is where a device called a *virtual audio cable* comes in. Picture it as connecting a cable to the "output" of SDRSharp and plugging it into the "input" of dsd—except that it's all happening virtually, completely in software. Two commonly used programs for Windows users are Virtual Audio Cable and VB Cable. Virtual Audio Cable has both a free and a paid version, while VB Cable is completely free.

Virtual Audio Cable can be found at http://software.muzychenko.net/eng/vac.htm, and VB Cable can be downloaded from https://www.vb-audio.com/Cable/index.htm. I experimented extensively with both programs, and while I had no luck at all with Virtual Audio Cable, I had no problems with VB Cable. I must specify here that I didn't try the paid version of Virtual Audio Cable, so it's possible that the paid version works just fine. Thus, your results may vary if you decide to try it, but my instructions going forward will be for VB Cable.

Once you've downloaded VB Cable, extract everything from the downloaded .zip file. Navigate inside the resulting folder and you should see, among a bunch of other files, a VBCABLESetup and a VBCABLESetupx64 application.

Choose the version right for your architecture, right-click it and choose Run as Administrator. All of the drivers should install, and if you check your Windows settings, you should see a new device, Cable Input, listed under your sound playback devices, and a Cable Output device under the recording tab.

While you have these settings open to check, set the VB-Audio Virtual Cable as your default recording device. This is because dsd—the program we're installing next—will automatically use the default recording device as its input.

Once VB Audio is installed, you'll need to download and install dsd. Ordinarily, this might be a tricky situation because the program is designed to run on Linux, and to run it on a Windows machine it needs to be compiled and installed using a Windows-based Linux emulator called Cygwin. Installing and running Cygwin can be instructive if you're interested in compiling and running dsd yourself, but it can also be problematic if you don't install it with all of the correct libraries, extensions, and compilers. It's a rabbit hole that many hobbyists may not want to follow.

We're fortunate that there are enough enthusiasts around that someone has done the hard work for us by compiling all of the necessary libraries and dsd itself, and then releasing the resulting Windows binary. This allows you to avoid the entire Cygwin-based rigamarole. You can download the binary from my github repo here: https://github.com/wdonat/jumpstarting_sdr.

Once you've downloaded the .zip file, extract the contents, which will give you a directory called dsd-1.7.0. Inside that folder you'll find the dsd application; don't open it just yet, as it's not something you just double-click and open. Just remember where you put it.

Now, open up SDRSharp and tune to a digitally encoded channel. In my experience, this can be one of the most problematic portions of the project, as there doesn't seem to be a central listing of broadcast frequencies—digital or otherwise—sorted by area. Try Googling your local police department and public safety organizations, or http://www.radioreference.com has a pretty comprehensive database. If all else fails, you may need to (as I did) simply start scrolling through the frequency dial and looking for a digital signal. They're pretty easy to distinguish, as they tend to be a digital "hum" sound and they're often broadcast either sporadically or in regularly spaced bursts. However, it's a large spectrum, and scrolling through it can take a long time, so you may want to save this option as a last resort. I wish there was a way to narrow it down, but all of my research up to now hasn't revealed a general area of the spectrum where these frequencies tend to reside. If you're aware of any such area, please let me know! I can say that in my area of southern California, I had my best results around 500 MHz. Obviously, your results may vary greatly.

Once you've found a digitally broadcasting candidate frequency, you'll need to adjust your SDRSharp settings. First, set the audio output to VB Audio. Then set the receiving mode to NFM (Narrow Band FM), and then set the bandwidth to about 12 KHz. When you've tuned to the strongest part of the signal, press the Play button.

Of course, you shouldn't hear anything, because instead of sending the output to your speakers as you did before, you're piping the output to your VB Audio installation, which in turn is piping to the dsd application (which hasn't started yet). Open a windows command prompt, and in that window, navigate to your extracted dsd-1.7.0 directory. Once you're inside that folder, enter

dsd -i /dev/dsp -o /dev/dsp -fa

This command tells dsd to listen to the default audio recording device, which you've set to be the VB Audio output, and pipe it to the default audio output device, which is most likely your speakers. Finally, the -fa flag tells it not to discriminate and to scan for all sorts of encodings.

If you've found a dsd-encoded frequency, the terminal window should begin scrolling text, which will change when speech is detected. The text scrolls (adds an additional line) every time the signal updates; for example, when a user presses their TALK button, that will show as a status update of text.

If nothing happens, there are two options: either the frequency you're listening to is not dsd-encoded or your settings are wrong. Double-check your settings and keep trying, including scanning the dial for likely signals.

One of the most important settings to play with is your gain—both in SDRSharp and in your Windows sound settings. Try different values in your SDRSharp output, and then open your Windows sound recording settings and adjust the microphone sensitivity. Once you've found a good digital speech frequency, it's unmistakable, and you'll really be able to hear a difference when you adjust either one of those settings.

Linux

Linux users will be using the same software as Windows users, but it's a bit more involved to build and install the program in Linux since there are no precompiled binaries to use.

That software is called dsd (for digital speech decoder), and in addition to your GQRX software, you'll need it and the packages it relies on. Windows users need to either install it using Cygwin or to download a pre-compiled binary, but Linux users will compile it for their particular platform. Start by visiting https://github.com/szechyjs/dsd and cloning the repository to your computer with the git clone command. Then visit https://github.com/szechyjs/mbelib and clone that repository as well.

Starting with the mbelib library, cd into the source folder and enter the following commands in your terminal:

```
mkdir build
```

```
cd build
```

```
cmake ..
```

```
make
```

```
sudo make install
```

This will install the mbelib application to your computer, which dsd depends on. When it's done, install the other needed precursors for dsd with the following command:

```
sudo apt-get install libitpp-dev libsndfile1-dev portaudio19-dev
```

When those are finished installing, go to the dsd directory in your terminal and type in the following:

```
mkdir build
```

```
cd build
```

```
cmake ..
```

```
make
```

```
sudo make install
```

That should install all of the necessary dependencies to your machine.

Now, unlike Windows, you will not need a virtual audio cable to pipe the output of GQRX to dsd. Instead, you'll need to use a program called Pulseaudio. It should be installed on most recent Linux distributions, but if it's not,

sudo apt-get install pavucontrol

will install it.

After installing these three applications—dsd, mbelib, and Pulseaudio, you're finally ready to start. Start up GQRX by typing gqrx in your terminal.

You'll need some specific settings to give dsd the sort of signal it can read, so set GQRX with the following parameters.

In the Receiver Options tab, set the mode to Narrow FM, choose the AGC setting of Fast, and set Squelch to about 60.

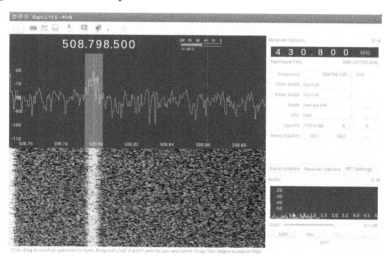

In the FFT Options tab, set your FFT size to 2048 and the sampling rate to 15 fps. You can increase these values if you have a particularly fast machine, but you shouldn't have any problems decoding with these settings.

Then open the options window in GQRX by clicking the settings icon (the one that looks like an IC just to the right of the Start button). Set your audio output to Default and click OK.

Finally, before you start GQRX, open another terminal window, navigate to the dsd/build/bin directory and enter the following command:

padsp -- dsd -i /dev/dsp -o /dev/dsp -fa -ma

This starts dsd, using as input (and output) the /dev/dsp device, which is a virtual device being written to by the Pulseaudio program. If you would like to see what other options are available to you, enter

./dsd -h

in your terminal to see the manual page.

Back in the dsd terminal, you should be greeted with the following text:

Digital Speech Decoder 1.7.0-dev (build:v1.6.0-xx-xxxxxx)

mbelib version 1.3.0

Audio In/Out Device: /dev/dsp

After that, the terminal window should remain empty, since you're not actually piping anything into it yet.

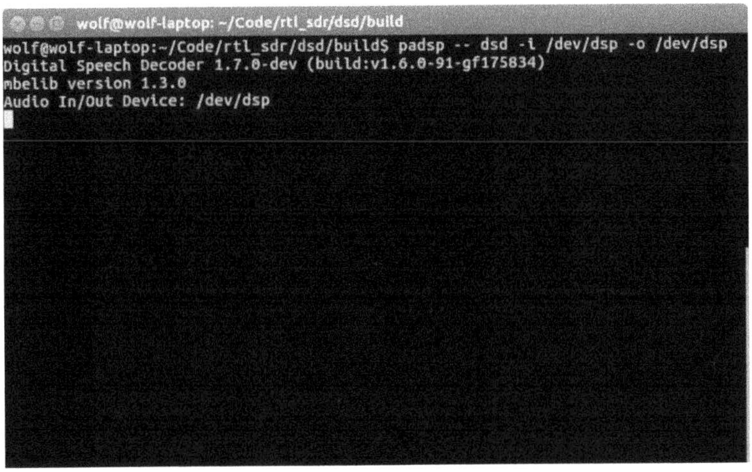

To start doing that, go back to your gqrx window. The main thing left to do is to find a digital voice signal, which is actually where I had the most trouble. You can search for dsd frequencies in your area, or http://www.radioreference.com has a pretty comprehensive database. If all else fails, you may need to (as I did) simply start scrolling through the frequency dial and looking for a digital signal. As you may have read previously in the Windows section, they're easy to distinguish, as they tend to be a digital "hum" sound and they're often broadcast either sporadically or in regularly spaced bursts. However, it's a large spectrum, and scrolling through it can take a long time, so you may want to save this option as a last resort.

You'll know when you find one because the terminal window in which dsd is running will suddenly start to scroll text as signals come in and are decoded.

It will, however, only scroll when digital voice data is being received; that is, if you're listening to an active channel, you're likely to hear only silence until someone actively presses the Send button on their radio. At that point, you'll see the word Voice in the dsd window and will hear the voice coming through your speakers.

Congratulations! You're now listening to digital voice signals, which should greatly expand your SDR horizons! Once you've found a channel, it becomes easier to find others, as you know what you're looking for.

For our final project, in the next chapter we'll track some weather satellites and grab some images sent over the airwaves.

Things to Try

1. Try to find all of your local government channels, including Fire, Police, Sheriff (if you have one), and even Forestry officials (so you can listen for forest fires in your area). Is there a particular band of the spectrum in which they tend to cluster?

2. See if there are digital speech channels within range of reception that are *not* affiliated with a government agency. Anything interesting? Pretend you're a private investigator; see what you can find out about any channels you come across.

CHAPTER 5

Listening to Satellites

Being able to tune into NOAA satellites while they pass overhead and download the images they're broadcasting is probably one of the coolest things about using an SDR setup. It's old-school technology that's been around since the 1960s but is now becoming easily accessible to hobbyists and enthusiasts, thanks to the SDR dongle and the software that you're now familiar with.

NOAA currently has quite a few weather satellites orbiting the Earth, keeping track of data worldwide, but there are only three that are broadcasting signals that are easy to receive. They broadcast around 137 MHz, and the signal can only be received when the satellite is passing overhead. Due to their orbital parameters, you can download the satellite's transmission and decode it about twice a day, depending on where you live.

The satellites—NOAA-15, 18, and 19—were launched in 1998, 2005, and 2009, respectively. They each broadcast a signal called Automatic Picture Transmission, or APT. APT was developed in the 1960s and is composed of two image channels, telemetry, and synchronization data, all transmitted as one horizontal scan line. You can use one of the programs we've already installed and used, GQRX or SDRSharp, to tune into each satellite's particular frequency. You can then either record the transmission and send the audio recording to another application for decoding or you can decode the transmission *on the fly* and display it as it's sent by the satellite overhead.

As it happens, just before this writing, NOAA released an official statement: "As of ~0000 UTC July 30, 2019 (DOY 211), the NOAA-15's AVHRR motor current has once again started spiking, becoming saturated above 302mA at ~0600 UTC. The instrument is once again no longer producing data and may be stalled." What this means for you is that if you happen to be trying to listen to 15's pass overhead and come up with nothing, give the other two satellites a try before giving up. I was able to tune into the satellite's signal, but I had no luck receiving an image, and I even got an error message I'd never seen before about missing telemetry data. So just be aware that NOAA-15 might be out of order when you read this.

The Doppler Effect

As a weather satellite passes overhead, you're dealing with the Doppler effect as it moves. What exactly is the Doppler effect?

The Doppler effect is the change in the frequency of a signal as you and the source move relative to each other. We experience it when we hear the siren on a moving ambulance rise in pitch as the vehicle speeds toward us, compressing the sound waves. If the ambulance doesn't hit us, we hear the siren decrease in pitch as the vehicle moves away, stretching the sound waves of the siren.

The same principle applies to light—and, as it happens, radio waves. Astronomers were first able to determine that certain stars and galaxies were moving away from us by seeing that the color of their light was shifted red. Likewise, a satellite moving at only 0.0025% the speed of light can have a frequency shift of a few kilohertz as it moves from horizon to overhead to horizon again.

We can even calculate the Doppler shift of each satellite, based on some easily researched numbers. NOAA 15 has an orbital semi-major axis of 4464.32 miles and an orbital period of 101 minutes. That works out to (4464.32 × 2 × pi)/(101/60) = 16,663 miles per hour. Call it 16,666 to make the math easier. The speed of light is 186,282 miles per second, which multiplied by 60 seconds and by 60 minutes equals 670,615,200 miles per hour. 1/(670,615,200/16,666) = 0.000025, or 0.0025 percent the speed of light. The Doppler equation is (c/c+v) × f. In this case that works out to (1/1.000025) × 137 MHz = 137,003,425, or a shift of 3.4 kilohertz. Speaking from experience, that's about the amount you'll have to adjust the tuning over the course of the satellite's pass overhead.

Hardware

The signal these satellites broadcast is *right-hand circular-polarized*, which means you won't be able to pick it up using the little antenna that came with your SDR dongle. A few antenna options are available to you, including a turnstile antenna, a quadrifilar helix antenna, or a V-dipole antenna. (For a more in-depth discussion of these and a few other antenna designs, make sure you check out the chapter on antenna theory and design.)

Although all of these options will work without any problems, it turns out that the easiest option to build and use is actually a set of old TV rabbit ears, mentioned earlier in the book.

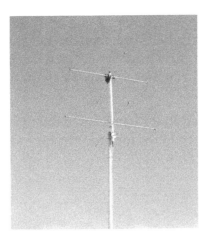

To pick up the APT transmissions using these rabbit ears, there are three requirements for the antenna that will most likely require you to modify those rabbit ears slightly.

1. The two antennas must be spaced 120 degrees apart. Most sets that are currently available only spread to about 90 degrees, so you'll have to make some small modifications. This may mean simply removing and remounting some screws so that you can spread the two antennas further apart. In some cases, you may have to cut or completely destroy the plastic base that holds the two antennas to spread them apart. Make sure that when you're remounting the two pieces, you use a protractor or something similar to get as close to a 120-degree angle between the two as you can.

2. The pair must be mounted horizontally rather than vertically. Again, this may mean simply mounting the plastic base on a vertical surface rather than a horizontal one, or you may have to use screws or adhesive to achieve it. It doesn't have to be perfectly flat—you don't have to use a level—but it does have to be close to horizontal.

3. Finally, the antenna structure must be oriented north-south. By that, I mean that if you picture the two antennas as the two parts of the capital letter V, and you lay the V horizontally, then the open end of the V should be pointing north. (Or, to look at it another way, the point of the V should point directly south.)

When it comes to placing your rabbit ears in the right orientation and direction, there are a few possibilities; my first attempt was simply some duct tape and a tall piece of PVC pipe.

It wasn't pretty, but it worked. My next attempt (shown in the next three images) was a bit more sophisticated—I used PVC again, but this time I also used a plastic flowerpot in which I cut out pieces of the plastic in order to set the antenna in it correctly. Finally, if you have access to some 3D-design software and a 3D printer you can probably come up with a fairly elegant solution. Luckily, the look of your design doesn't matter, only the results.

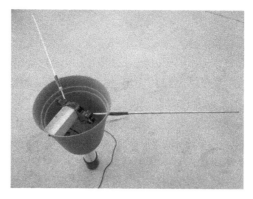

And some other views...

Once you've built your antenna, you have to find where to place it. All the antenna needs is a clear view of the sky, so a backyard or a rooftop is probably going to be your best bet. Again, this is where the cable extension and adapters I spoke about in Chapter 3 will come in handy, as it's nice to be able to place the antenna in your backyard and be able to work on the software portion of things in your living room.

Once you've got your antenna set up, it's time to configure the software. Unlike the signal decoding in the digital speech chapter, I had equal success receiving and decoding NOAA signals with both Linux and Windows.

Software

As always, I've divided the software section into Windows and Linux. In both operating systems, you'll be using another freeware program called WXtoImg to convert the received signals to images. Unfortunately, the developer who created it suddenly stopped supporting and updating it and sort of vanished, but a group of enthusiasts managed to resurrect the code and the site so that it may continue to be used. Point your browser to https://wxtoimgrestored.xyz/downloads/ and get the correct version for your platform.

Windows

In Windows, you'll also be using your SDRSharp program as well as VB Audio again to pipe the output to your WXtoImg software.

Start by getting the WXtoImg application. The installation package for Windows works fine on the latest Windows 10.

Once it's installed, double-click to open it and you should see a blank screen with multiple tabs: Image, Audio Files, Raw Images, Saved Images, Composites, and Animation.

The first thing you need to do is to tell the program where you're located. Click on Options -> Ground Station Location and enter either your city name or your latitude and longitude if you happen to know it. When that's done, go to Options -> Update Keplers to download the latest updates of the satellite orbital parameters. This will tell you what satellites will be passing in range, when, and for how long. When the Kepler updates have finished downloading, click on Options -> Satellite Pass List, and you should see a screen like the one in the picture on page 45.

Choose a satellite from the three available; it doesn't really matter which one, as your results should be the same. When you have a satellite picked out, take note of the time it will be passing overhead, and its broadcast frequency.

Next, go to Options -> Recording Options and set the soundcard to CABLE Output (VB-Audio), and make sure the sample rate is set to 11025. Now go to Options -> Auto Processing Options. Make sure Record and Auto Process is selected, and check the Create Image(s) box.

Finally, go to File -> Record and click the Auto Record button at the bottom of the screen. This will activate WXtoImg automatically when the next satellite is in range. This is a pretty nice feature of the program that I think is overlooked; you can schedule the recording so that you don't have to worry about starting the program.

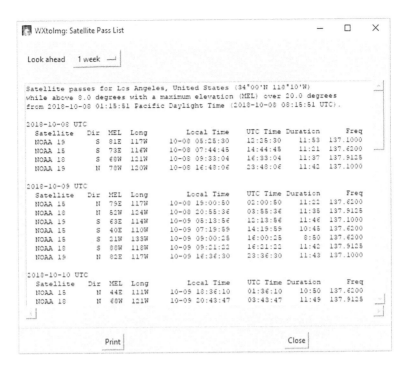

That completes the WXtoImg setup, so let's open SDRSharp. Tune to the frequency at which your chosen satellite will be broadcasting. In the Radio tab, set the Receive mode to WFM and the Bandwidth to about 37 KHz. In the Audio tab below that, make sure that Filter Audio is unchecked. Finally, set the Output in that same tab to CABLE Input (VB-Audio Virtual Cable) or the closest alternative you have in your menu.

That finishes setup. It seems like a lot of settings, and you may be tempted to skip one or two, but *don't do that.* All of these settings have been carefully curated by SDR priests and priestesses, guarded closely and handed down among the faithful. I now pass them off to you.

Now you just have to wait until the satellite's pass-over time. When you're about a minute or so away from that time, start up SDRSharp and start looking for the satellite's broadcasting signal. Make sure your frequency matches the one listed in the satellite pass list, but be prepared to adjust it quickly and on the fly; the cheaper SDR dongles can sometimes be 10 s of KHz out of adjustment, so a satellite broadcasting at 137.1 MHz may show up as 137.2 or 136.9 when the signal starts to come through. You're looking for a tightly grouped batch of peaks on the frequency graph. At first the display will look like the first image shown on page 46, but as you zoom in and the satellite comes into range, it'll look like the second picture shown on page 46.

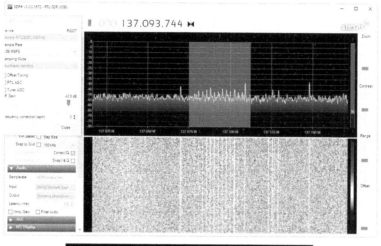

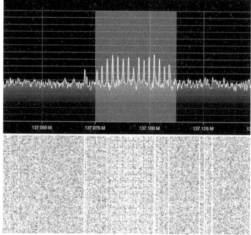

At the same time, WXtoImg will come to life and start recording, and you should see it filling in the image, line by line. Keep adjusting the gain control in SDRSharp, and keep actively tuning as the pass progresses. You want to keep the tuning bar right in the center of the peaks, and this center will move because the signal will drift due to the Doppler effect as the satellite passes overhead.

When it's done, your WXtoImg window should show something like the first image on page 47, which is the raw capture.

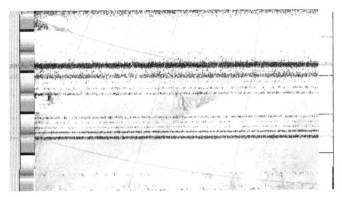

Now, allow WXtoImg to perform its automatic processing, and then click the Saved Images tab. With any luck, you should have something like this:

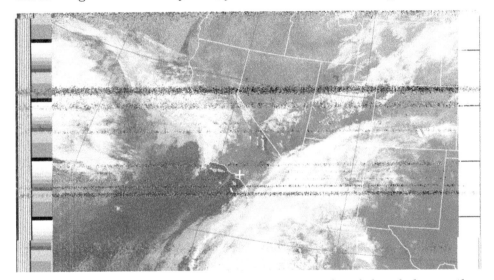

If so, congratulations! You've successfully captured and decoded a weather satellite image!

Linux

Like Windows, Linux uses a freeware application called WXtoImg to convert received signals to images. I used the .deb package for my Ubuntu 16.04 installation and had no problems with using it; just double-click the file and allow Ubuntu to open and install it with the native software installation application.

Once it's installed, open the program by entering

xwxtoimg

in a terminal window (it's the GUI version of the software, which is much easier to use). You'll be greeted by a blank screen with multiple tabs: Image, Audio Files, Raw Images, Saved Images, Composites, and Animation.

The first thing you need to do is to tell the program where you're located. Click Options -> Ground Station Location and enter either your city name or your latitude and longitude if you happen to know it. When that's done, go to File -> Update Keplers to download the latest updates of the satellite orbital parameters. This will tell you what satellites will be passing in range, when, and for how long. When the Kepler updates have finished downloading, click File -> Satellite Pass List, and you'll be greeted by a list of satellites that are passing, when they'll be overhead, how long the pass will last, and what frequency they're broadcasting on.

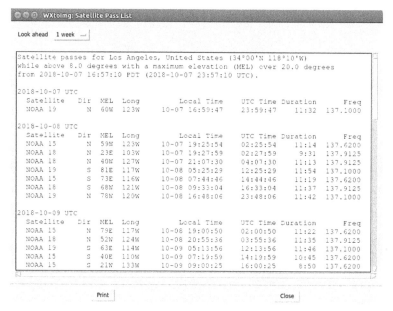

Now you're ready to open Gqrx from a terminal with

gqrx

Choose a satellite from the three available; it doesn't really matter which one, as your results should be the same. When you have a satellite picked out, take note of the time when it will be passing overhead and its frequency.

You'll have some settings to adjust in order to receive the satellite's automatic picture transmission (APT) signal correctly. In the Input Controls tab, turn off Hardware AGC, Swap I/Q, and No Limits, and turn on DC Remove and IQ Balance. Set LNA to about 20 dB to start, but be prepared to revisit this, because I had to increase this to around 40 to hear the signal when it passed. Set your frequency correction to 0.

In the Receiver Options tab, choose Normal for both Filter Width and Filter Shape, set the mode to Narrow FM, and AGC to Fast.

In the FFT Settings tab, set your FFT size to 2048 and the rate to 15 fps, which should be plenty fast enough to catch the signal without taxing your processor, even if it's a slow one.

Set your tuning frequency to the one listed in the satellite pass list for your chosen satellite, but be prepared to adjust it; the SDR dongles can sometimes be 10 s of KHz out of adjustment, so a satellite broadcasting at 137.1 MHz may show up as 137.2 or 136.9 when it starts to come through.

Lastly, in Audio Options, select a recording folder by clicking the button with three dots to the right of the UDP, Rec, and Play buttons. Click the Recording tab and select a folder. This way you can click Rec when the satellite begins its pass. The sound file will be placed in the directory you choose, and you can analyze it later if WXtoImg doesn't work.

That finishes your GQRX setup, so go back to WXtoImg. Click Options -> Recording Options and make sure that the soundcard selected is Default Audio and the sample rate is at 11025.

Now click Options -> Auto Processing Options and make sure that Record and Auto Process is selected, along with Create Image(s), (as shown on page 50), which will let the software create the appropriate images as the satellite passes.

WXtoImg: Record

Record only (show image if enabled)

• Record and auto process

    ✓ Create image(s)   Image Settings...

     Create movie(s)   Movie Settings...

     Create composite image(s)   Composite Image Settings...

     Add images to web page   Web Page Settings...

Minimum percent of projection filled   0.1

Minimum solar elevation for visible images   15.0

Exclude from composite images and animations

Minimum Good Quality Scan Lines   0 (0:00)

Show MCIR when MSA fails on screen

Remove audio files   never

Remove raw images   never

Remove maps   never

Remove images   never

OK   Cancel

That completes your setup. It may seem like a lot of settings, and you may be tempted to skip one or two, but *don't do that.* All of these settings have been carefully curated by SDR priests and priestesses, guarded closely and handed down among the faithful. I now pass them off to you.

Finally, to get ready, go to File -> Record and click the Auto Record button at the bottom. WXtoImg will now wait until the next scheduled satellite pass; at that time, it'll start recording and processing the incoming data. This is a pretty nice feature of the program that I think is overlooked; you can schedule the recording so that you don't have to worry about starting the program while you're worried about the time, whether your antenna is pointed correctly, or any number of other issues to worry about. At least you'll know that your image-processing software is coming online at the right time.

All you have to do, then, is start the Gqrx feed a minute or two before the pass is scheduled, and make sure your frequency is correct and that you're receiving. On Linux, you can pipe the output of the sound to your speakers; it's pretty easy to hear when you've locked onto the signal—it's a high-pitched beeping noise. On the frequency plot, it'll look like a tightly grouped patch of symmetric peaks broadcasting regularly.

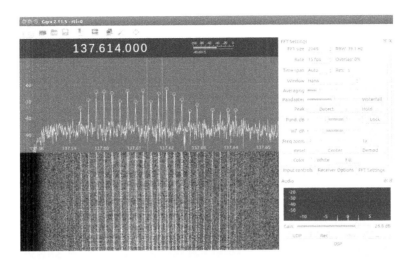

Your WXtoImg software should flicker to life (possibly before you even recognize the signal) and begin showing the image as it is transmitted, line by line. Of course, it'll be nothing but static at first, but an image will slowly appear as the signal gets stronger.

Back in Gqrx, you should be staying busy. Keep your tuning bar in the middle of the patch of peaks, and be prepared to change the tuning, because the Doppler effect will move the frequency of the signal as the satellite passes. Keep playing with the Gain slider as well. The bottom right corner of the WXtoImg window has a Volume indicator showing the strength of the signal coming in; you want it to be green, showing 50 percent signal strength. As the picture comes into focus, you'll start to pick up on what to adjust and how it affects the picture.

If all goes well, you should end up with a picture like this in your picture WXtoImg window.

The static bands will come and go as you play with the tuning. When the satellite is finished passing overhead, WXtoImg will stop recording automatically and begin processing the images it received. Wait until it's finished, and then check the Saved Images tab. You should see a host of different images, color-corrected and cleaned up for you, including one that may look similar to this:

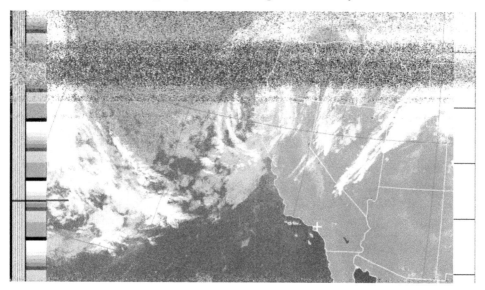

If so, congratulations! You've successfully received and decoded your first NOAA satellite image!

Troubleshooting

Don't be too frustrated or disappointed if it doesn't work well the first time. Check the satellite pass list and prepare for the next pass, which may come anywhere from two to twelve hours later. It becomes easier to recognize the signal each time you see it, and when you know what you're looking for, it becomes easier to tune to it as well. You'll get better each time you try it, and you may end up with some truly stunning results. You may consider switching satellites as well; you may have different results on a different frequency, and since the orbital parameters differ, you may not have to wait that long for the next pass of a different satellite. It didn't take me long to know exactly what I was looking for, signal-wise, and I started watching for it like a hawk when the pass time got close. If you're like me, you'll definitely experience a thrill as the signal starts to coalesce, and it's pretty darn cool when WXtoImg kicks on and you start to see an image slowly appear.

Try This

1. The International Space Station transmits radio signals at 145.80 MHz. Do some online research to determine when the ISS will be passing over your location and see if you can use one of your existing antenna setups to tune in. http://www.n2yo.com and http://www.heavens-above.com both have tools that allow you to track the ISS and determine when it will next pass over your current location.

2. Pulsars emit radio waves as well, often around 400MHz. Is it possible to use your setup to capture the signal from a pulsar? How much amplification would you need to use? What would it even sound like?

3. If, after reading this book, you're hooked on the possibilities inherent in using SDR to listen to radio frequencies, try stepping it up in terms of the quality of the equipment you're using. SDRPlay is a unit that many people have had success with, costs around $100, and comes with some good-quality software to play with.

Conclusion

I hope this book has given you some idea of the incredible number of possibilities that SDR offers to the casual hobbyist. For just a few dollars in parts, you can investigate all parts of the radio frequency spectrum and even decode some truly beautiful images from space.

It can be a little confusing to get started, but once you do, I think it's well worth the time and brain power spent figuring everything out. Good luck with your reception, and please feel free to send me some of your results. I can be reached at wolframdonat@gmail.com or on twitter: @wolfram_donat. Happy tuning!

Running SDR on the Raspberry Pi

After playing with all of this stuff, you probably started thinking, "Why can't I make this portable? What prevents me from creating an on-the-go package that lets me take my SDR setup to the middle of Death Valley and seeing what RF signals I can pick up?" Or even "Why can't I make an SDR boombox?"

Well, dear reader, I've got you covered. As it happens, GQRX will run on the Raspberry Pi, and I've gone through the headaches and dependency-chasing rigamarole for you. I now offer to you the hidden knowledge of how to get your SDR setup working on a Raspberry Pi. Add the official 7-inch touchscreen and a battery pack, and you're ready to take your setup anywhere on the planet.

As I write this, the Pi v4 has been out for a few months. I haven't tested this with the very latest 64-bit OS, but everything you're reading from here forward has been tested to work successfully on both a Pi 3B+ and a first-edition (32-bit) Pi 4. The Pi 3B+ was running Stretch, and the Pi 4 was running the Buster version of the Raspberry Pi OS with a kernel version of at least 4.14. (To check, run

uname -a

and

cat /etc/os-release.)

As you can probably imagine, getting SDR to work on the Pi is almost identical to getting it to work on any other Linux installation, especially Ubuntu, since Rasbpian is a Debian derivative. It turns out that the hardest part of getting GQRX to run on the Pi is chasing dependencies and making sure you don't have conflicts between installed programs. GQRX was created and is maintained by Alexandru Csete, and he has released a package that should run on your Pi.

To start with, create a folder for your SDR experiments with

mkdir SDR

Then, cd into the SDR directory and download everything from here:

https://github.com/csete/gqrx/releases/download/v2.11.5/gqrx-sdr-2.11.5-linux-rpi3.tar.xz

The easiest way to do this, in my opinion, is to use wget, but you can also enter that link into your browser's address bar.

Once it's downloaded, unzip the .tar using

tar xvf gqrx-sdr-2.11.5-linux-rpi3.tar.xz

and then cd into the resulting folder.

The readme.txt in the directory is very helpful, but my experience was that it wasn't complete. It's possible that because this is an older build of GQRX, it requires dependencies that aren't installed by default in newer versions of Raspbian. To make sure you've got everything you need, do the following in your terminal:

sudo apt-get install cmake build-essential git

sudo apt-get install gnuradio gnuradio-dev

sudo apt-get install qt5-default qttools5-dev qttools5-dev-tools qtmultimedia5-dev

sudo apt-get install libqt5svg5-dev libqt5webkit5-dev libsdl2-dev libasound2 libxmu-dev

sudo apt-get install pavucontrol libportaudio2

sudo apt-get install libvolk1-bin libusb-1.0.0 gr-iqbal

Then, following the instructions in the readme, from the gqrx-sdr, enter

sudo cp udev/*.rules /etc/udev/rules.d/

to enable your Pi to access USB devices.

Next, just as in Ubuntu, you'll need to see if the DVB-T drivers are loaded, though they probably are. Enter

sudo rmmod dvb_usb_rtl28xxu

in your terminal. If you don't get a response, that means that the drivers were loaded and your command was successful. In that case, you'll need to disable the drivers permanently. Using sudo, create the file /etc/modprobe.d/rtlsdr.conf

In that file, enter

blacklist dvb_usb_rtl28xxu

Save the file and reboot.

Finally, you'll definitely want to run the volk optimizations. Volk (which is Linux for "vector optimized library of kernels") is a library profiler. It tests various software on your computer to find the versions that best match your processor:

volk_profile

It'll take a while to run, but the optimizations it performs should make a difference in your Pi's SDR performance. Now you should be ready. Just like the Ubuntu setup earlier, connect your antenna to your dongle, and then plug your dongle into a free USB port on the Pi. (Remember, I'm using the Pi 3B+ for this, which has only USB 2.0 ports. If you're trying this using the Pi 4, I don't know how things will work using the USB 3.0 ports. If you try them and nothing works, try a USB 2.0 port before starting over.)

To start GQRX, open a file explorer window and navigate to the gqrx-sdr-2.11.5-linux-rpi3. Double-click the gqrx icon.

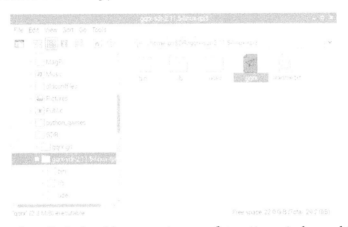

The first window that should pop up is a configuration window asking you to configure the I/O devices, as shown on page 58. Select the Realtek RTL2838 device from the first drop-down menu, leave the rest as defaults, and click OK.

You should be greeted by the same gqrx screen as you were earlier in Ubuntu. To test it, tune to a local FM radio station. Change the Mode to WFM(mono) and click the triangular Play button at the top left. You may need to plug a pair of headphones into the Pi's audio jack to hear the station, as there are a few bug reports that the gqrx audio doesn't play well through the HDMI connection. My headphone setup worked well, and I was listening to my local station.

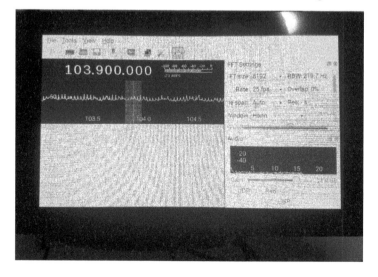

I/Q input

Device: Complex Sampled (IQ)

Device string: 0,repeat=true,throttle=true

Input rate:

Decimation: None

Sample rate: 0 Msps

Bandwidth: 0.000000 MHz

LNB LO: 0.000000 MHz

Audio output

Device: Default

Sample rate: 48 kHz

Cancel   OK

The last step will be hooking up your touchscreen, if you're so inclined. I tried it and it opens without any problems. However, I did discover immediately that the reception from the antenna was quite a bit worse, though nothing had changed in my setup, antenna-wise. It's pretty obvious that the touchscreen itself is putting out a lot of noise, and this is picked up by the SDR dongle. If you want to create a boombox, you may need to experiment with a different antenna, or perhaps even move the touchscreen away from the Pi (my layout has the Pi directly connected to the screen using the standoffs).

If you think about it, it's both really cool and ironic that you can go through all of these steps and end up with a (gasp!) handheld radio that lets you listen to local stations. *If only they'd come up with something like that a long time ago...*

Try This

Some things to try with your Pi setup:

1. I joked about it, but try making a boombox! You can use a small LiPo (Lithium Polymer) battery to power the Pi and your setup, as long as you also use a voltage regulator (the Pi has no onboard regulator). Experiment with different placements of all of the ingredients—the Pi, the screen, and the power supply—to see how reception is affected.

2. If you're feeling particularly confident, the Pi 4 will run Ubuntu 19.04 and 19.10. See if you can get the Linux programs mentioned in this book to work on the Pi after installing the necessary libraries and dependencies, rather than using the prepackaged version we used in this chapter. (Hint: make sure you're looking for ARM-compatible libraries).

Thank you!

How did you enjoy this book? Please let us know. Take a moment and email us at support@pragprog.com with your feedback. Tell us your story and you could win free ebooks. Please use the subject line "Book Feedback."

Ready for your next great Pragmatic Bookshelf book? Come on over to https://pragprog.com and use the coupon code BUYANOTHER2021 to save 30% on your next ebook.

Void where prohibited, restricted, or otherwise unwelcome. Do not use ebooks near water. If rash persists, see a doctor. Doesn't apply to *The Pragmatic Programmer* ebook because it's older than the Pragmatic Bookshelf itself. Side effects may include increased knowledge and skill, increased marketability, and deep satisfaction. Increase dosage regularly.

And thank you for your continued support,

The Pragmatic Bookshelf

Design and Build Great Web APIs

APIs are transforming the business world at an increasing pace. Gain the essential skills needed to quickly design, build, and deploy quality web APIs that are robust, reliable, and resilient. Go from initial design through prototyping and implementation to deployment of mission-critical APIs for your organization. Test, secure, and deploy your API with confidence and avoid the "release into production" panic. Tackle just about any API challenge with more than a dozen open-source utilities and common programming patterns you can apply right away.

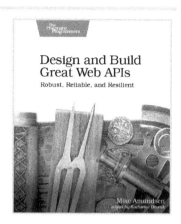

Mike Amundsen
(330 pages) ISBN: 9781680506808. $45.95
https://pragprog.com/book/maapis

Quantum Computing

You've heard that quantum computing is going to change the world. Now you can check it out for yourself. Learn how quantum computing works, and write programs that run on the IBM Q quantum computer, one of the world's first functioning quantum computers. Develop your intuition to apply quantum concepts for challenging computational tasks. Write programs to trigger quantum effects and speed up finding the right solution for your problem. Get your hands on the future of computing today.

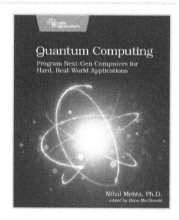

Nihal Mehta, Ph.D.
(580 pages) ISBN: 9781680507201. $45.95
https://pragprog.com/book/nmquantum

A Common-Sense Guide to Data Structures and Algorithms, Second Edition

If you thought that data structures and algorithms were all just theory, you're missing out on what they can do for your code. Learn to use Big O Notation to make your code run faster by orders of magnitude. Choose from data structures such as hash tables, trees, and graphs to increase your code's efficiency exponentially. With simple language and clear diagrams, this book makes this complex topic accessible, no matter your background. This new edition features practice exercises in every chapter, and new chapters on topics such as dynamic programming and heaps and tries. Get the hands-on info you need to master data structures and algorithms for your day-to-day work.

Jay Wengrow
(506 pages) ISBN: 9781680507225. $45.95
https://pragprog.com/book/jwdsal2

Build Location-Based Projects for iOS

Coding is awesome. So is being outside. With location-based iOS apps, you can combine the two for an enhanced outdoor experience. Use Swift to create your own apps that use GPS data, read sensor data from your iPhone, draw on maps, automate with geofences, and store augmented reality world maps. You'll have a great time without even noticing that you're learning. And even better, each of the projects is designed to be extended and eventually submitted to the App Store. Explore, share, and have fun.

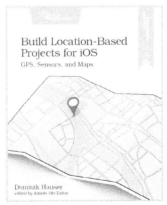

Dominik Hauser
(154 pages) ISBN: 9781680507812. $26.95
https://pragprog.com/book/dhios

iOS Unit Testing by Example

Fearlessly change the design of your iOS code with solid unit tests. Use Xcode's built-in test framework XCTest and Swift to get rapid feedback on all your code — including legacy code. Learn the tricks and techniques of testing all iOS code, especially view controllers (UIViewControllers), which are critical to iOS apps. Learn to isolate and replace dependencies in legacy code written without tests. Practice safe refactoring that makes these tests possible, and watch all your changes get verified quickly and automatically. Make even the boldest code changes with complete confidence.

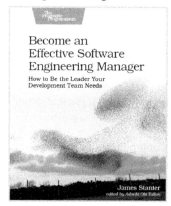

Jon Reid
(358 pages) ISBN: 9781680506815. $47.95
https://pragprog.com/book/jrlegios

Become an Effective Software Engineering Manager

Software startups make global headlines every day. As technology companies succeed and grow, so do their engineering departments. In your career, you'll may suddenly get the opportunity to lead teams: to become a manager. But this is often uncharted territory. How do you decide whether this career move is right for you? And if you do, what do you need to learn to succeed? Where do you start? How do you know that you're doing it right? What does "it" even mean? And isn't management a dirty word? This book will share the secrets you need to know to manage engineers successfully.

James Stanier
(396 pages) ISBN: 9781680507249. $45.95
https://pragprog.com/book/jsengman

Build Websites with Hugo

Rediscover how fun web development can be with Hugo, the static site generator and web framework that lets you build content sites quickly, using the skills you already have. Design layouts with HTML and share common components across pages. Create Markdown templates that let you create new content quickly. Consume and generate JSON, enhance layouts with logic, and generate a site that works on any platform with no runtime dependencies or database. Hugo gives you everything you need to build your next content site and have fun doing it.

Brian P. Hogan
(154 pages) ISBN: 9781680507263. $26.95
https://pragprog.com/book/bhhugo

Practical Microservices

MVC and CRUD make software easier to write, but harder to change. Microservice-based architectures can help even the smallest of projects remain agile in the long term, but most tutorials meander in theory or completely miss the point of what it means to be microservice based. Roll up your sleeves with real projects and learn the most important concepts of evented architectures. You'll have your own deployable, testable project and a direction for where to go next.

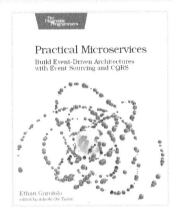

Ethan Garofolo
(290 pages) ISBN: 9781680506457. $45.95
https://pragprog.com/book/egmicro

The Pragmatic Bookshelf

The Pragmatic Bookshelf features books written by professional developers for professional developers. The titles continue the well-known Pragmatic Programmer style and continue to garner awards and rave reviews. As development gets more and more difficult, the Pragmatic Programmers will be there with more titles and products to help you stay on top of your game.

Visit Us Online

This Book's Home Page
https://pragprog.com/book/wdradio
Source code from this book, errata, and other resources. Come give us feedback, too!

Keep Up to Date
https://pragprog.com
Join our announcement mailing list (low volume) or follow us on twitter @pragprog for new titles, sales, coupons, hot tips, and more.

New and Noteworthy
https://pragprog.com/news
Check out the latest pragmatic developments, new titles and other offerings.

Save on the ebook

Save on the ebook versions of this title. Owning the paper version of this book entitles you to purchase the electronic versions at a terrific discount.

PDFs are great for carrying around on your laptop—they are hyperlinked, have color, and are fully searchable. Most titles are also available for the iPhone and iPod touch, Amazon Kindle, and other popular e-book readers.

Send a copy of your receipt to support@pragprog.com and we'll provide you with a discount coupon.

Contact Us

Online Orders:	*https://pragprog.com/catalog*
Customer Service:	*support@pragprog.com*
International Rights:	*translations@pragprog.com*
Academic Use:	*academic@pragprog.com*
Write for Us:	*http://write-for-us.pragprog.com*
Or Call:	+1 800-699-7764

Milton Keynes UK
Ingram Content Group UK Ltd.
UKHW031815121223
434254UK00008B/697